COMPUTING
FOR
BEGINNERS

THE BASICS
EXPLAINED
IN PLAIN ENGLISH

By Lynn Manning

Bloomington, IN Milton Keynes, UK

authorHOUSE

AuthorHouse™
1663 Liberty Drive, Suite 200
Bloomington, IN 47403
www.authorhouse.com
Phone: 1-800-839-8640

AuthorHouse™ UK Ltd.
500 Avebury Boulevard
Central Milton Keynes, MK9 2BE
www.authorhouse.co.uk
Phone: 08001974150

First published by AuthorHouse 4/4/2006

ISBN: 1-4259-1542-6 (sc)

Printed in the United States of America
Bloomington, Indiana

This book is printed on acid-free paper.

"Microsoft product screen shots reprinted with permission from Microsoft Corporation".

CONTENTS

INTRODUCTION

So why another book about computing – and why was I the one to write it?

It came from my personal experience........ in 1989 I was struck down with ME and was forced to leave my job as a secretary, unable to work for 10 years and completely losing my confidence. And of course, by the time I came to work again, computers had taken over the workplace – and I knew nothing about them.

So in 2000, anxious about returning directly to work as a mother of a teenage son and a computer ignoramus, I decided to re-train and enrolled on a beginner's computer course at my local Adult Education Centre. Fortunately for me, the course began by explaining how to switch the computer on!

By the time I finished my training in 2002 I had completed more computing courses and had more certificates than you could shake a stick at ... City & Guild Certificates in Basic Computing, Internet, E-mail, Word-processing, Spreadsheets, Desktop Publishing and also a Certificate in Business and Office Technology. You name it, I had it.

But there was a problem with these courses. More than one problem – learning in a group has its disadvantages. One person picks things up quickly and is able to progress – while others have difficulty grasping the subject and find themselves struggling. Many people find they attend courses but can't put into practice what they have learnt because their computer or software is different. This gave me my business idea. My plan to return to an office job went out the window – instead I started my computer training business.

So in 2002 I set up Computing for Beginners which was to offer one to one computer training. With one to one training you are more relaxed, feel able to ask questions and can generally progress quicker.

Only recently, several of my customers have suggested I write a book saying I should write it in the same manner in which I teach them. That means plain English, no computer jargon, just an easy-to-follow manual for the beginner building on my own experience as a computing latecomer, now training newcomers!

So, that's where the book came from and how I came to write it – one thing I would suggest is that as a beginner you read the introductory chapters thoroughly before you progress onto the chapters dealing with the Internet, E-mail, Word-processing. Now let's get started......

YOUR COMPUTER

Your computer system will consist mainly of four components, the tower, the monitor (screen), the keyboard and the mouse. You may also have a printer.

Before I go any further, there is one important thing to remember, your computer is a machine – you are in charge!!!

There are several words you will come across which may cause some confusion so I am therefore going to quickly explain some of these words before moving on…..

HARDWARE: The main components of your computer (tower, monitor, keyboard, mouse, printer) are known as hardware (basically hardware refers to components which you can see). All of these components work together and are all connected to the tower via cables.

HARD-DRIVE (C: Drive): This is the main storage in your computer and is located inside your tower so you cannot actually see this.

CD/DVD DRIVE: This refers to one of the slots on the front of your tower into which you can place CDs or DVDs either to load a program, to view a film or to save whatever you have been working on.

SOFTWARE: To enable a computer to function it must use software (software is the program with which you do something with your computer and usually comes on a CD. It is the information on the CD which is the software). The CD would be inserted into one of the slots on the front of your tower and this would enable the relevant program to be installed onto your computer. The CD would then be taken out and kept safe. Once the software has been installed, it will stay on your computer's hard-drive. Good examples of software are your email program, Internet Explorer (which is a program which you will use to search the internet), and your word processor.

OPERATING SYSTEM: The most important piece of software on your computer is your operating system and if you have just purchased a new computer you will be using the latest version of Microsoft Windows which at the moment is XP. It is the operating system which makes your computer work. It is called "Windows" because every time you open a new file or program, it will open in a new window. The window can fill the whole screen or be minimised to one third of its size. Windows are always rectangular.

****

THE TOWER

The *TOWER* is actually your computer, in effect like the brain, which translates all of your instructions and collects information, without which nothing else works. A lot of people believe it is the monitor or screen which is the computer, which is an easy mistake to make as it is the monitor that you are looking at when using the computer.

It is on the tower that the *power* button is located. This is the button which turns the computer on (known as booting up a computer). There may also be a *reset button* which will enable you to restart your computer without completely turning it off (known as *rebooting*). You may need to re-start your computer should everything *freeze* but don't worry this wont happen too often. Also on the front of the tower there will be slots (drives) into which CDs/DVDs or floppy discs are placed. These enable you to load programs onto your computer, play CDs or DVDs or to save your work. There will also be a light on the tower which indicates that the computer is on.

THE MONITOR

The *MONITOR* or screen is where you will see all the images and the text that you type in. When you first switch on you will see a small arrow in the centre of your screen (controlled by your mouse) which is called a *cursor*.

THE KEYBOARD

The *KEYBOARD* is used to type in text and you can also use the keyboard to move around the screen. The main keys on the keyboard are arranged in the same way as on a typewriter. Some of the keys you will not be using for some time but because they are on the keyboard, I will explain what they all do.

Unless you are using a laptop computer, your keyboard
will look very similar to the picture above.

You will see at the top of your keyboard a row of keys labelled
<u>F1 to F12</u>. These keys are called *Function keys* but generally
they are not used. The exceptions being the F1 key which will
bring up a help menu in any program and F4 which when
pressed with the Alt key will close the current window.

The Print Screen, Scroll Lock and Pause Break keys, which are
located next to the function keys, are almost redundant and you
probably won't use these keys. These were designed for 1980s
software and are rarely used by modern Windows programs.
The *Print Screen* key enables you to take a copy of the screen
you are viewing and save it onto a clipboard. The *Scroll Lock*
key is used mainly in Excel which is a spreadsheet program
and it enables you to move the window without moving the
selected cell. The *Pause Break* key, when pressed at the same
time as the *Windows logo* key will bring up all the information
about your computer. As I say I don't think you will need to
use any of these keys at the moment so don't worry too much
about them.

The *Escape* key can be used to stop things happening. You
can use this in Internet Explorer as it has the same effect as
pressing 'cancel' or 'stop'.

The *tab* key can be used to move several spaces to the right or to move to the next option on a page.

The *caps lock* key will enable everything to be typed in capital letters (remember to release the key afterwards or you will continue to type in capital letters!)

The *shift* key, when held down will type a capital letter and when released will revert back to normal type. You will notice that some keys have multiple characters on them – if you hold down the shift key at the same time as pressing one of these keys you will type the character at the top of that key. You have two shift keys, one at each end of the keyboard.

The *Control* key *(Ctrl)* when pressed with another key will enable you to take various shortcuts instead of using the toolbars. Depending on which program you are in at the time, Ctrl *and S* will save your work and *Ctrl and P* will print your work.

The *Windows logo key*, when pressed, will bring up the Start menu.

The *Alt* key is used for various shortcuts i.e. Alt + F4 will close the window currently open.

The *Spacebar* is used to put a space between words.

The *Arrow* keys enable you to move around a page. The up and down arrows scroll the page up and down. These arrow keys, when used in a word-processing program will move the cursor around.

The *Enter* key takes you to the next line when typing. It is also pressed at the end of an instruction to tell the computer to carry out that instruction.

The *Backspace and Delete* keys are used to delete the last letter you have just typed. Press Backspace once and it will delete the last letter typed to the left of the cursor and by pressing Delete you will delete the last letter typed to the right of the cursor. Backspace can also be used when on the Internet to take you back to the previous page (same as pressing the back button).

The *Insert* key is used in word-processing to overtype everything to the right of the cursor. Simply click on the Insert button and whatever you type will replace the words already there. Don't forget to click on it again to revert to normal typing.

Home and End keys. These keys can be used in conjunction with the Control key. For example if you press Control and home, you will immediately be taken to the top of your document and similarly if you press Control and End you will be taken to the end of your document.

The *Page Up and Page Down* keys do just that. Pressing page up will move the page up and similarly page down will move the page down.

Num Lock key. This key has two uses. If you press this key so the function is on, you can type numbers from the small panel on the right of your keyboard. If you press it again and take the function off, the keys on the number pad can be used to scroll around your page.

Num Indicator light and Caps Indicator light. If the number indicator light is on then you can use the number pad on the right of the keyboard to type the numbers. If the Caps indicator light is on, this is telling you that you have selected to type in capital letters.

Holding down *Ctrl, Alt and Delete* all at the same time will enable you to end a program which may have frozen.

THE MOUSE

The *MOUSE* itself has a few different functions.

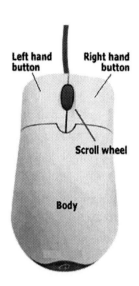

Before we can begin you need to know how to use your mouse. The mouse is a device which you use to point to things on the screen. When you move the mouse you will see a small arrow (*cursor*) on the computer screen. You use the cursor to point to the object or program you wish to select. The cursor must be on the object or program required.

It is a good idea to place your hand over the mouse so you feel comfortable and relaxed and so that your index finger on your right hand (if you are right-handed) is resting over the left hand button on the mouse. This way you won't have to move your whole hand when you need to click – just your finger. Don't worry if you are having difficulty – everyone has trouble at first controlling the mouse – just try and relax your hand.

Some phrases which you may come across are:

Cursor: This is a symbol, like your finger which points to information on your screen or shows your position on the screen. It can be a pointer, a hand, cross hairs, a blinking vertical line, an hourglass, or a number of graphics but typically one of these.

Left Click:	this means click once on the left button – this is the most often used function and is used to give the computer an instruction.
Double Click:	click twice, in quick succession, on the left button. This is used to open a program or folder.
Right Click:	Click once on the right hand button. This will bring up a mini menu specific to the item clicked on.
Scroll:	Most mice have a scroll wheel which you can use to move the page up and down.
Drag:	This involves holding down the left side of the mouse and moving the mouse in order to drag and drop files into folders.

GETTING STARTED

This chapter will be split into different sub-sections as follows:

1. Switching on and what to expect
2. The Desktop, and the Icons
3. The Taskbar, Start Button and finding Programs
4. My Documents
    a. Files and Folders
    b. Making a new Folder
    c. Moving a file into a Folder
    d. Deleting a file or a Folder
    e. Renaming a file or Folder
5. Recycle Bin
6. Windows (resizing and scrolling)

1. Switching on and what to expect

So we are ready to go and you have pushed the power button on your computer; the computer screen will flicker on and off and you will see several different images one of which is a black screen with white writing on it, which will disappear as quickly as it appeared – don't worry this is all perfectly normal and is just the computer setting itself up. You may be tempted to panic now but don't, just be patient and wait. The computer will settle down in a minute!

2. The Desktop, and the Icons

The Desktop

The screen you will see when the computer finally settles is called *"the Desktop"*. This screen will always be there, even when you open a new window the desktop will always be running underneath.

Before we go any further let me just recap about the term windows. Windows is your operating system (a program which makes your computer work). It uses rectangular areas of the screen called "windows" to display information.

Icons

The above is a picture of what may appear on your "desktop". You will see various little pictures (called icons) which represent shortcuts to programs installed on your computer. You may have many more icons on your desktop relating to additional programs which are installed on your computer.

It is a good idea to only have icons on your desktop which relate to the programs you use most often. If you think of your desktop in the same way you would think of an office desk – that way you will only place the things on it which are used regularly. Otherwise you will find the desktop getting cluttered and you won't be able to find things quickly.

In order to access any of these programs from the icons on your desktop, you will need to double click on them. This in effect will open the program. Once you are in a program all instructions only require a single click of the mouse.

The icons on the desktop are *"My Documents"* which is a folder into which all your work will be saved on your computer.

"Recycle Bin" into which everything you delete from your computer is stored. The things in your Recycle Bin can be restored at any time and are not completely deleted until you empty the Recycle Bin.

Internet Explorer. This is the program you will use to "surf" the internet.

Outlook Express. This is the most commonly used e-mail program.

Microsoft Word. This is a program used for word-processing (writing letters etc).

3. The Taskbar, Start Button, Finding Programs and Turning off your Computer

The Taskbar

At the bottom of the desktop you will see a bar. This is called the **"Taskbar"**.
On the right of the taskbar is the clock and also programs which are running in the background but which need not concern you.

The taskbar is used to quickly switch from program to program or window to window. This bar will display all programs currently open. On the picture below, there are four programs currently open. To switch between programs, simply click once on the program required on the taskbar.

This bar also contains the **"Start"** button on the left hand side.

The Start Button

By clicking once on this button, you will be given access to several windows functions. These include shutting down the computer, accessing the help and search functions, the Control Panel, and most importantly all programs installed on your computer can be found here. The picture shown on the next page is what you will see when the Start button is clicked.

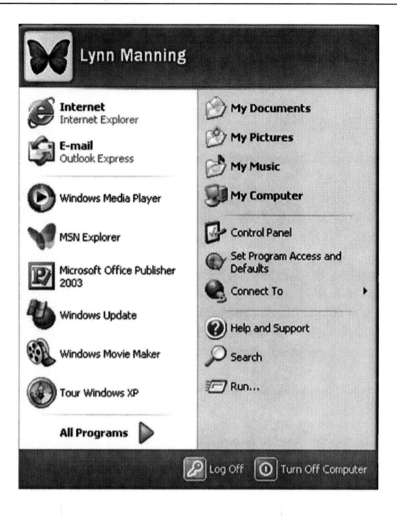

Your name will be displayed at the top, and on the left hand side of this small window will be the programs you have used most recently with Internet and Email always being at the top. The list at the bottom on the left will change according to which programs you have recently used. To use any of the programs from here, just left click once on the program icon required and it will open.

The list on the right hand side will never change and displayed here are items used regularly such as My Documents, My Pictures and My Computer. Below these you will find Control Panel, which you should not need to use at this point but you should know that from here you can change the settings on your computer and remove certain programs. Below this are help, search and run and these basically do what they say.

Finding Programs

It is always a good idea when you have just bought a new computer to spend a few minutes going down the programs list so you can see exactly what programs are installed on your computer.

To find out which programs are installed on your computer, click once on the Start button and then click once on the button at the bottom with the green arrow next to it called **"All Programs"** (see picture shown on next page) and you will see all the programs installed on your computer. If there is a program title on the list which has a black arrow next to it, this means there is a sub-menu attached to it and if you just position your mouse on one of these (no need to click) a sub-menu will appear showing more programs.

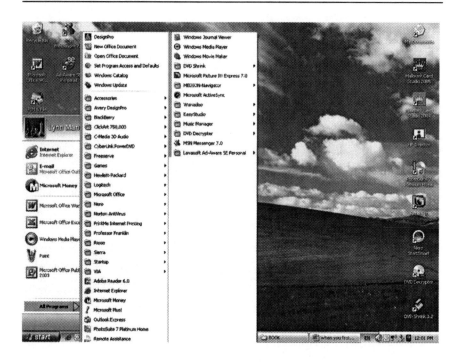

If you find a program you would like to open, just click once on it and it will open.

Turning Off your Computer

At the bottom of the start menu is the Turn off Computer button. When this is pressed a small window will appear in the centre of your screen (see picture shown on next page). You then need to click on the Turn off button and your computer will turn itself off. **You should always turn the computer off this way and never turn it off on the tower.** There is usually a short delay before the computer shuts down.

4. My Documents

"My Documents" is a folder on your computer into which most of the work you complete and wish to keep, whatever program you are working in, is usually saved to. This folder is purely for storing information and not for the preparation of work. All letters you may write in the word processing program (which will be covered later) will be saved into My Documents. There are a few other folders which you will see in "My Documents". These are "My Pictures" which in time, if you are using a digital camera, will hold all your photographs which you transfer to your computer. Another folder which you will see is "My Music". Again any music which you transfer to your computer either by copying from a cd or downloading from the internet will be stored in this folder. There are two other folders which you will see in "My Documents" which are "My e-Books" and "My Videos". Neither of these are important at the moment.

a. Files and Folders

When you create a piece of work (letters etc) which you wish to keep, these are called files and are saved in the "My Documents" folder. You can have as many files in "My Documents" as you like but it is always a good idea to keep things tidy. Just as "My Documents" is a folder and will hold all your files, it is perhaps better to create separate folders in My Documents which relate to different subjects i.e. a folder for business letters, and a separate folder for personal letters. That way if you need to find any of your letters and you have filed them correctly, you will know straight away where to look. Again, you can have as many folders as you like – there is no limit. The same principal will apply to the "My Pictures" folder.

b. Making a new Folder

If you have saved several files and wish to organise and keep them together then you should make a new folder.

When you open up "My Documents" you will see a pane on the left hand side of the window which contains three separate boxes. Click once on "Make a new folder" in the top box. Once you have done this, you will see a folder symbol in the "My Documents" window appear with a black rectangle containing the words "New Folder". You can now name this folder whatever you like, i.e. "Business Letters" by overtyping.

18

c. Moving a file into a Folder (drag and drop)

Now you have made your new folder, you need to move your file into the folder and the easiest way to do this is to drag and drop it into the folder. To do this you need to position your cursor over the file to be moved, hold down the left button on the mouse, and keep it held down, and move the file over the top of the folder and then let go. Don't worry you haven't lost the file; it's just gone into the folder. To check it is there, double click on the folder and the file will be inside – it's all very clever.

d. Deleting a file or Folder

You may have files or folders in "My Documents" which are old and which you no longer wish to keep. If this is the case, you should delete them by doing the following.

Click once on the relevant folder so it appears highlighted.

Folder to be deleted

Click on "Delete this Folder" in the pane on the left of the window.

Click Yes on the small window which has now appeared in the centre of the screen (see picture shown on next page) and this will send the file or folder to the Recycle Bin.

File and Folder Tasks

Rename this folder
Move this folder
Copy this folder
Publish this folder to the Web
Share this folder
E-mail this folder's files
Delete this folder

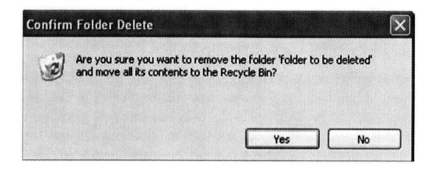

Everything in the Recycle Bin can be recovered. The items are not completely deleted from your computer until you actually empty the Recycle Bin.

e. Renaming a file or Folder

If you wish to change the name of a file or folder, you should do the following:

Click once on the relevant file or folder so it appears highlighted.

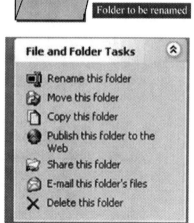

Click on Rename this folder in the pane on the left of the window.

Rename the folder by overtyping.

5. Recycle Bin

When you delete something from your computer, it is placed in the Recycle Bin. It is not completely deleted from your computer until you empty the bin. To do this, double click on the Recycle Bin icon on your desktop and you will see all the items you have recently deleted.

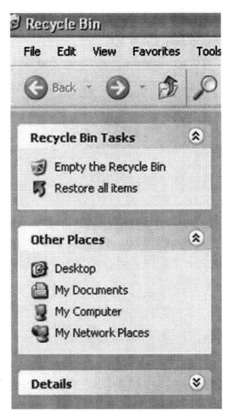

To empty the bin, click once on "Empty the Recycle Bin" in the pane on the left of the window.

If you have deleted an item which you wish to restore to its original destination, then click on the relevant item, then click on "Restore this Item" again in the pane on the left of the window.

6. Windows (resizing and scrolling)

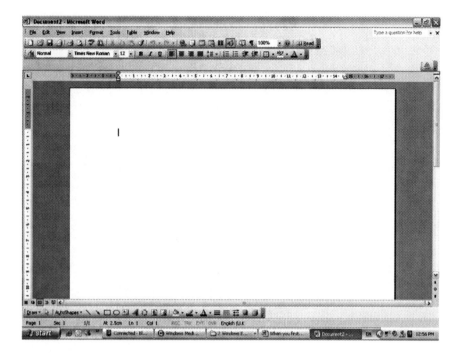

Above is a picture of a window which I have opened in Microsoft Word which is a word processing program. Windows are the same whatever program you are in and the above is just an example.

Resizing Windows

On the top right hand corner of every window you will see three buttons. ▬▣✖ By clicking once on ▬ which, is the minimize button, this will appear to close the window altogether but actually what you have done is to put it on the taskbar below so it is still open but out of the way. To get the window back, just click once on the program displayed on the taskbar and it will be restored to full size.

By clicking on 🖻, which is the restore down button, the window you are working on, which at the moment is filling the whole screen and is full size, will resize itself down to about a third of its actual size. You will now notice that this button has changed to 🔲 which is the maximise button. By clicking on this, you will restore the window back to its full size. Sometimes when you open a new window, it won't be full size so you will have to click on the maximise button to make it fill the screen. The ❎ is the button which will close the window altogether and this button is always red with a white X on it. This is the button you always press to exit the program you are working on.

If you are on the internet and press on the ❎ button, the window will close. However, if you have been working in your word processor or similar program and you click on the ❎ a small window will appear asking you if you want to save the changes you have made. This is a good thing, because this is a good reminder that you may not have saved your work.

Scrolling with Windows

On the right hand side of every window is a scroll bar with small arrows either end. By clicking on the ▲ arrow this will scroll the page up. Similarly by clicking on the ▼ arrow you can scroll down the page. By using either of these arrows the page will move up or down one click at a time. You can also scroll the page up or down by placing your cursor over the bar between the two arrows and holding down the left button on the mouse and then moving it up or down.

Some windows also have a scroll bar at the bottom and again the same principal will apply. The ◄ and ► arrows will move the window to the left or right and again the bar can be moved in the same way as above.

You can also use the scroll wheel in the centre of your mouse to move the page up and down, if applicable.

So now you should know the basics of your computer and where things are. The next chapters will deal with specific programs and will take you through getting the most out of them. We will start with the Internet.

THE INTERNET

Getting Online

In order for your computer to dial up to the internet you must have a connection established with an internet provider. You may have a dial up connection on pay as you go or anytime or you may have a broadband connection. The difference between these is that on a pay as you go connection you will pay only for the time your computer is dialled up to the internet and the cost is usually the same as a local phone call. These charges will appear on your phone bill. If you have an anytime connection, you will pay a flat monthly fee to your internet provider, via direct debit, and you will be able to use the internet at any time. You cannot receive phone calls if you are using a dial up connection because your computer is using your phone line to establish the connection. If, however, you have a broadband connection then this allows you to use the internet and your phone at the same time as special filters are supplied for you to place in your phone socket which, in effect, split your phone line, so half can be used by your computer and the other half by your phone. Again with broadband you will pay a flat monthly rate by direct debit. There is a significant difference between a dial up connection and a broadband connection and that is the speed at which you will be able to use the internet. Broadband can be up to 20 times faster than a dial up connection.

Internet Explorer explained

Internet Explorer is the program you will use to explore the internet and this program is known as your browser. Every computer will have this program installed and it will be represented by the icon as shown on the previous page.

Internet Explorer will enable you to visit websites and find out information. There are websites about all sorts of subjects and most organisations will have one. Websites are used to do your shopping online, find out the latest news and weather, play games, and find information ... in fact almost anything.

You will have to use your mouse to move around the internet pages. You simply move your mouse and this controls the pointer on the screen. You will have to click the left side of the mouse to issue an instruction or choose an option. This is usually an arrow ↖ but will change to I when you point it at text or where you wish to type text but when you have your cursor over a part of the page and it changes to a pointing finger ☝, this means that you can click here and you will be taken to a new page of your choice.

To begin using the internet you will need to double click on the Internet Explorer icon on your desktop – if you are using a dial up connection a small window will now appear and you must click on connect. Your computer will now establish a connection with the internet. If, however, you are using a broadband connection then you will need to click on the broadband connection icon on your desktop **before** clicking on the Internet Explorer icon.

The first screen you will see will be your Internet provider's home page. On the picture shown on the next page you will see that my Internet provider is Wanadoo and it is their home page which opens every time I access the internet. Think of this page as your base from which you can begin to explore the internet.

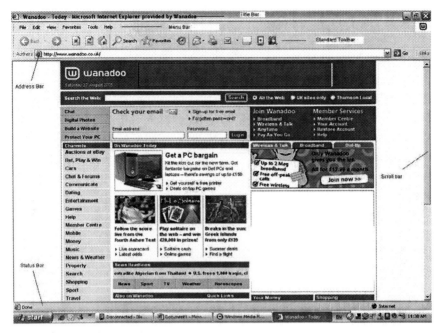

First I will explain the Internet Explorer window. At the top of the page is the title bar which tells you which website you are on and this bar also has the minimize, maximise and close buttons at the right hand side. It is the white cross in the far right hand corner which you will need to click on to close the window when you have finished.

The next line down is called the menu bar which contains File, Edit, View, Favourites, Tools, and Help. We will leave these for now as you won't need to access these yet.

The next bar down is called the standard toolbar which contains icons which you will use a lot and these are explained, in detail, on the following pages.

Standard Toolbar

Back – Forward buttons

The first two icons on this bar are back and forward. By clicking once on the back button you will be taken back to the page previously viewed and likewise forward will move you forward. These buttons will only be activated (they will be dark green) when you have viewed several pages. You can also move back a page by clicking on backspace on your keyboard.

Stop button

The next icon is stop. If a webpage is taking a long time to load or you have changed your mind about visiting this site you can click on this button to terminate the instruction.

Refresh button

The next icon is Refresh. If the webpage being viewed has not properly loaded then you can click on this button to reload the page.

Home button

The next icon is home. If you are feeling a bit lost and wish to return to somewhere familiar then by clicking on this button, you will return to the homepage.

Search button

The next icon is Search. This is typically used if you do not know the website (www) address. If you need to search for something you can click on this button and a panel will open on the left hand side so that you can type in the item you are searching for. Your homepage will also have a search box into which you can type your subject.

Favourites button

The next icon is Favourites. Favourites are where you will store the addresses of frequently used websites. That way you will not have to keep typing in the www address every time you wish to visit that particular website. To add a website to your favourites you should do the following:

1. Get onto the website required.
2. Click on Favourites (a panel on the left will appear)
3. Click on Add (at the top of this panel)
4. Click on OK on the "Add Favourite" window which now appears.

You will now see the relevant website has been added to the list in Favourites.

Next time you wish to visit this site, just click on the name in the Favourites list and you will be taken straight to the website.

History

The next icon is History. By clicking once on this icon you will be able to see all websites visited over the past 2 weeks.

Email

By clicking on the next icon your email program will open.

Print

The next icon is the print button. By clicking on this button you will be able to print the page you are viewing. Sometimes when you print from the internet the right hand side of the page does

portrait

landscape

not print properly so you need to change the page from portrait to landscape. To do this you will need to click once on the word File on the menu bar, then click on page setup and then in the small window which now appears click in the small circle next to landscape then click ok.

The above are the only buttons on the standard toolbar which you will be using at the moment.

Address Bar

The next bar down with the white box is called the *address bar* and it is in this box that you type the www address of the website you wish to see. So if you know the address of the website, you must first click in this box and then type in the address. You will then need to click on *Go* to the right of the address bar or press the enter key on your keyboard and you will be taken to that website.

Remember you can only use this if you know the exact website (www) address. Otherwise use Search.

Status Bar

The bar just beneath the web page window at the bottom is called the *status bar* and you can see on here whether the page has loaded correctly. When you have typed in a website address or clicked on search to search for something, the page will start to load and at the bottom of the page, on the *status bar,* you will see a small rectangular box which will gradually fill with green bars. When the box is full, the page has loaded. This is a good indication of what is happening as sometimes you may think nothing is happening so keep your eye on this box.

Scroll Bar

The bar on the right hand side of the screen is called the scroll bar and you can use this bar to move the page up and down. (You can also use your page up and page down buttons on your keyboard to do the same) Sometimes you will have a scroll bar at the bottom of the page as well, and this can be used to move the page from left to right.

Finding a Website

First of all double click on the Internet Explorer icon and wait until your homepage appears. Then in the address bar at the top of the page, type in the address of the website required (don't forget to click first so that you can begin typing) i.e. www.computingforbeginners.co.uk. You will now need to click on Go to the right of this bar or you can just press the enter key on your keyboard and you will be taken to the website. The website will now appear as per the picture below.

When you first get on to a new website, the first page you will see will be the homepage of that particular site. Every website will have a list of links (either at the top or the side of the page) which will take you to different parts of that website and some have links to different websites.

You will see on my website that I have four links at the top of the page which are Home, About Us, Services, and Contact Information. By clicking on any of these you will be taken to the relevant section of the website.

If you just move your cursor, which at the moment is an arrow, so it is over the "About Us" heading you will notice that the cursor has now changed to a pointing finger and if you click now you will be taken to the page entitled "About Us".

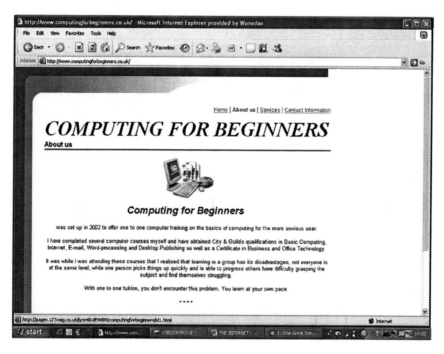

Again if you now go back to the top and click on "Home" you will be taken back to the first page of this website. It is basically just like turning the pages of a book. As long as the cursor is a pointing finger when you click you will be taken to a different page of the site.

Searching for Information

To make a search for a specific item you must type in a word or words in the search box, for example in the picture below I have searched for "flights to Italy" and have used the search box on my homepage to do this search.

You now have the option to search UK sites only or the whole web. Click in the circle beside the required option and then click on the search button to the right of the search box. The results of the search are shown in a list as per the picture on the next page. When you do a search for something you always get a long list of results, sometimes thousands of results but I always find it's the top 10-15 results which are the most relevant. To select one of the results, you should click on the one you want to look at (again make sure your cursor is a pointing finger before you click) and you will be taken to the website of your choice.

Be as specific with your keywords as you can and you will get more accurate results.

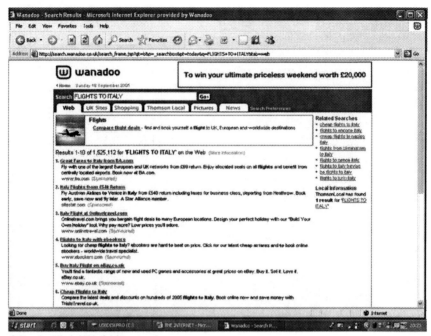

E-MAIL

E-mail, which is short for electronic mail, is a quick and easy way to keep in touch with other people. Your messages are sent and received instantly over your phone line. You can send e-mails to someone the other side of the world and it will be received in an instant. To be able to use e-mail you must have an e-mail address which you would have been given when signing up with your internet provider.

The most common e-mail program is Outlook Express and this will already be installed on your computer. To begin using this program you must first double click on the Outlook Express icon on your desktop or if this is not on your desktop, you should  click on Start, then All Programs, find Outlook Express in the list and click once on it and the program will open.

The picture on the next page is an example of what you will see when you open Outlook Express.

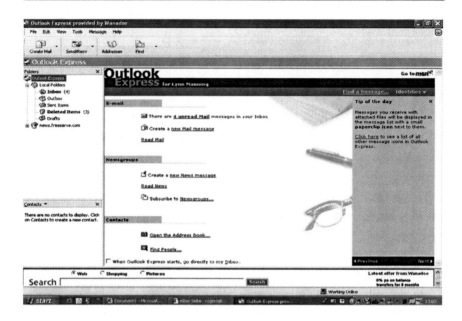

First I will explain the above window. At the top of the page is the title bar on which you will see Outlook Express.

Menu Bar

The next bar down is your menu bar which contains File, Edit, View, Tools, Message and Help. You will not be using a lot of the functions in these menus at the moment. Those you will need, I will explain below.

File - by clicking once on this menu, you will be able to save and open messages, create new folders to store messages, import and export messages from another email program and create a new identity which will enable two different people to use the same Outlook Express program.

Edit – by clicking once on this menu, you will be able to delete email, move messages to folders, and empty your Deleted Items folder as well as other editing tasks.

View - by clicking once on this menu, you will be able to change the way you view your messages and also change the entire look of Outlook Express.

Tools - by clicking once on this menu, you will be able to send and receive emails, change settings and access your address book.

Message – this menu contains most of the functions which are available on your standard toolbar, such as reply, reply to all, forward and you can also block senders from this menu.

Help - by clicking once on this menu, you can access help menus relevant to Outlook Express.

Standard Toolbar

The next bar is the toolbar which contains icons which you will be using a lot so I will briefly explain what they do as I shall go into greater detail further on in this chapter.

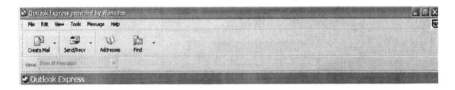

Create Mail, will when clicked bring up a new message window to enable you to compose and send a new e-mail message.

Send/Receive, will when clicked send any messages which are in your outbox and at the same time check for any new messages and, if there are any will bring them into your Inbox.

Addresses, will when clicked bring up your address book and you can put all your email addresses in here.

Find, will enable you to find an email quickly. You can search for an email by inserting the name of the sender, the subject of the message or a specific word.

Folders

You will see that there is a panel on the left hand side of the Outlook Express window which is split into two, the top half contains all your folders; Inbox, Outbox, Sent Items, Deleted Items and Drafts. Any new folders which you make will also appear in this list.

Inbox into which all new messages will be placed.

Outbox which will hold messages waiting to be sent.

Sent Items which will contain copies of all emails you have sent.

Deleted Items is the folder into which all messages which have been deleted, from whichever folder, are placed. These messages will stay in this folder until you empty it.

Drafts is a folder into which incomplete messages can be stored – to be finished at a later time.

The bottom half of this pane contains all your contacts. Contacts are the people you have placed in your address book.

Sending an E-Mail

To send an e-mail you must first click on Create Mail which is the first icon on the toolbar at the top of the page. The following New Message window will then appear in the centre of the screen.

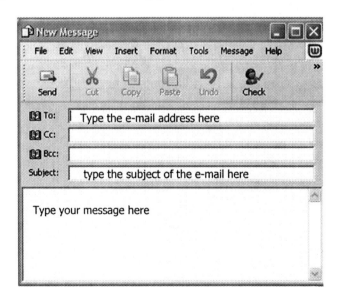

First of all you must click in the To: box at the top and insert the email address of the person you are sending the message to i.e. mary@hotmail.co.uk. (be very careful when entering e-mail addresses as a small mistake will result in the email not being sent.) You should then click in the subject box and insert a subject relevant to the e-mail i.e. if the email is regarding your holiday, you would type holiday in this box.

You should then click anywhere in the large white box at the bottom and begin to type your message. Your message can be as short or as long as you want. When you have finished typing your message you should then click once on the Send button (first icon on the toolbar). Your email will now be sent and will be received in a matter of minutes, wherever in the world it has been sent. It's as easy as that.

You can check that this message has been sent by clicking once on the Sent Items folder on the left hand side of the window. You will then see instantly the date and time your message was sent.

Receiving an E-Mail

To check for any new messages, you should click once on the Send/Receive button at the top of the window. If you have any new messages these will now be placed in your Inbox and if you look at the top part of the main page, it will say **"you have 1 unread message in your Inbox"**.

Usually when you open Outlook Express it will automatically check for new messages.

Read your E-Mail

To read your messages you should click on Read Mail or click on Inbox, both of these actions will take you to your Inbox where all incoming mail is delivered. You will notice that the main part of this window is split into two. The top half will tell you who the email is from and the subject of the message. If you click once on the email in this top half, so it is highlighted, the message can be seen in the preview pane at the bottom of the window. You may not be able to see the entire message in the preview pane and may have to use the scroll bar to the right of the page to view the whole message. Alternatively, if you double click on the email, in the top part of the window, the email will open in its own window making it easier to view. Click on the maximise button to enlarge the window and when you have finished reading the message, click on the cross in the top right corner and you will be taken back to your Inbox.

Replying to an E-Mail

If you have received an e-mail which you wish to reply to you should click on the Reply button on the standard toolbar and type your message where the cursor is flashing. Then click on Send. There is no need to insert an e-mail address as this message will go straight back to the address from where it has been sent.

Deleting E-Mails

Once you have read your messages and you no longer wish to keep them, you should click once on the email to highlight it and then click once on the Delete button at the top of the page represented by a big red cross. By doing this, the email has now been placed in your Deleted Items Folder. It will not be totally deleted from your computer until you empty your Deleted Items folder.

Emptying Deleted Items Folder

Once you are sure that all the messages which have been placed in your Deleted Items folder are no longer required, you should empty this folder. To do this, click once on Edit on the menu bar at the top, then click once on "Empty Deleted Items Folder". You will then be asked if you are sure you wish to delete these messages, click yes if you are sure. Remember when you have emptied this folder, you cannot get the messages back so be absolutely certain before doing this.

Making a new E-Mail Folder

You may have received an email which you have read and which you do not wish to delete but which you wish to keep in a separate folder. For instance you may receive a lot of emails from your friends and family so you could make a new folder entitled "Friends and Family".

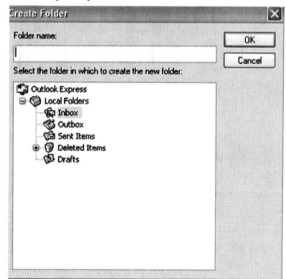

To do this you should first click on File on the menu bar at the top of the window, then bring your cursor down to folder (no need to click) and a sub-menu will appear. Move the cursor over and click on new. You will now see a "Create Folder" window. Type in the name of the new folder in the Folder name box at the top of this window and then click on the - (minus) next to Local Folders and then click OK. Your new folder will now appear in your folder list on the left hand side of the Outlook Express window.

Moving E-Mails to Folders

To move emails to folders, you should first click on the email you wish to move so it is highlighted, then click on Edit on the menu bar at the top and then click on 'move to folder'. You will now see a "Move" window. Click once on the folder into which you wish to move your email and then click on ok. Your message will now have been moved to your chosen folder.

Sending Attachments

You may wish to send an attachment with your email i.e. a text file or a photo. There are two different ways of doing this. The first way is to click on **Create Mail** and type your email message as explained above, and then click on **Insert** on the menu bar at the top of the window and click once on **File Attachment.** (*You can also click once on the paperclip symbol which appears on your standard toolbar*) You will now see an "Insert Attachment" window.

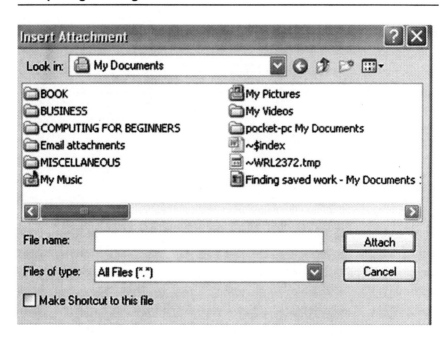

This is where you select the location of the file to send. To the right of the "Look In" box is a downward arrow. If you click on this arrow a list of locations appears. Most files are kept in "my Documents" so you would click on this in the list and then click once on the file to be attached. Click once on attach at the bottom of this small window and your file is now attached to your email. Just click the send button to send your email with your attachment.

The other way and perhaps the better way if you are sending pictures is to come out of Outlook Express and go into "My Pictures". Locate the picture you wish to send, click once on it and then on the left hand side in the "File and Folders Tasks" pane, click once on "e-mail this file". You will now see a small window entitled "Send pictures via e-mail".

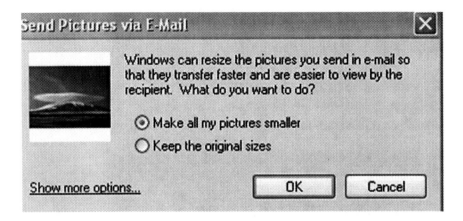

This is basically asking whether you want Windows to resize your picture in order for it to be sent quicker. Click Ok on this window, after first checking that "make all my pictures smaller" option has been selected.

A new e-mail window will now appear with your picture attached. All you now have to do is to insert the e-mail address of the person you are sending the message to and type your message. Simply click Send after this and the message and your attachment will be sent.

Opening an Attachment

If an email has an attachment with it, you will see a small paperclip symbol next to the email in the top half of the window. You will also see a paperclip symbol in the preview pane below, on the right hand side. To open the attachment you should click on the paperclip symbol in the preview pane and then click on the name of the attachment. You will now be asked if you want to open this file, click open and the attachment will open.

Address Book

To add new contacts to your address book, you should first click on Addresses at the top of the window. The following window will appear.

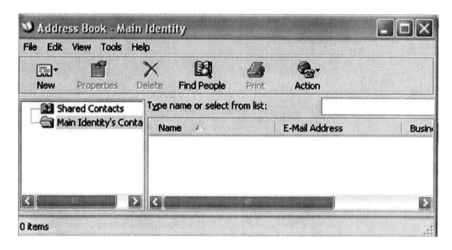

You should now click once on New and then click once on New Contact where you will see the following window.

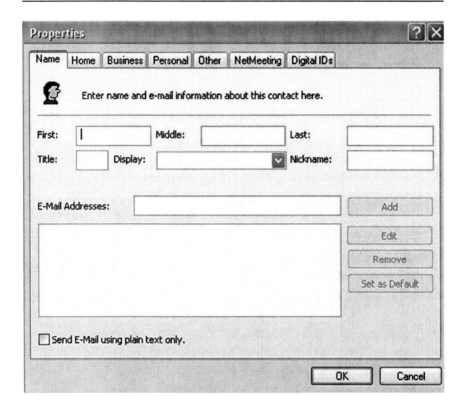

You should type the first name and last name of the contact into the relevant boxes and then type in the email address. You will see there are various tabs at the top of this window enabling you to enter all sorts of information such as home address, telephone numbers, business information, personal information etc. You need only insert as much information as you want but be sure to click on OK after all required details have been inserted. This will then save this contact to your address book and they will appear in your contacts list, in the bottom left pane of the Outlook Express main window.

You can send an email to a person in your contact list by double clicking on the required name in this list and a new message window will appear with the email address already inserted. Simply type a subject, type your message and send.

WORD PROCESSING

The most commonly used program for word processing is Microsoft Word and I have based this chapter on this program. There are other programs you may be using such as Microsoft Works Word Processor which is very similar. Unless an icon for this program has been placed on your desktop as a shortcut, you will need to click once on Start, then hold the cursor over All Programs and find Microsoft Word in the program list. Click once on it and you will have opened Word. If it is on your desktop, simply double click on the icon.

Depending on which version of Word you have installed on your computer, the first window you will see will be very similar to the one shown below.

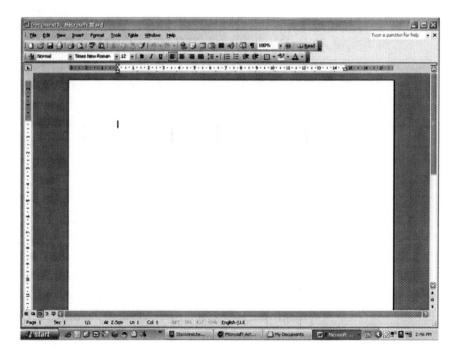

At the very top of this window is the **Title Menu**, which, when you start a new document will always say Document 1 – Microsoft Word and its on this bar that the minimise, maximise and close buttons are found on the right hand side.

Menu Bar

The next toolbar is called the **Menu Bar** which consists of File, Edit, View, Insert, Format, Tools, Table, Window, and Help. By clicking once on any of these you will be presented with sub-menus enabling you to choose a number of commands which I will explain in detail. I will explain only the items you will be using and will ignore the items which are more advanced.

File – By clicking once on this you will be presented with the following menu which will among other things, allow you to open, close, save and print your document.

New will open a new blank document.

Open will open a document which you have previously saved.

Close will close the current document you are working on.

Save will save any changes which you have made to your document.

Save As allows you to choose where you want to save your document and allows you to give a name to your document.

Page Setup allows you to change your margins and the general layout of your page.

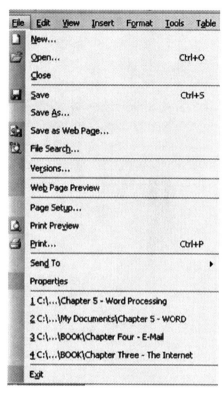

Print Preview allows you to view your document on the screen before you actually print it out on paper.

Print allows you to print your document.

Edit - by clicking once on this you will be able to carry out editing tasks such as cut, copy and paste.

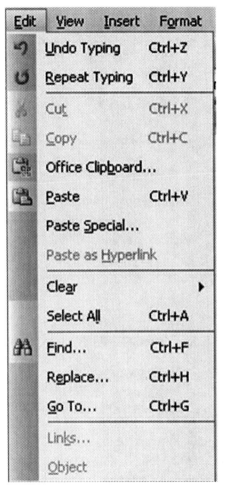

Undo Typing will quickly erase the last thing you have typed.

Redo Typing will quickly retype something you may have erased by using undo typing.

Cut allows you to take out a piece of text from your document.

Copy allows you to copy an item of text and place it in a different place on your document or you could place it in a new document.

Paste allows you to put the 'Cut' or 'Copied' text into a new location in your document.

Select All allows you to highlight all the text in the document you are working on.

**View** – by clicking once on this you will be able to change the way you view the current window, and add or remove toolbars.

Normal, Web Layout, Print Layout and Reading Layout allows you to choose how you wish to view your document on screen.

Toolbars allows you to select which toolbars you have showing on your screen.

Ruler allows you to show the ruler at the top and left hand side of the page.

Header and Footer allows you to add information to the top and bottom of your document. For instance you could have the date at top of every page and page numbers at the bottom.

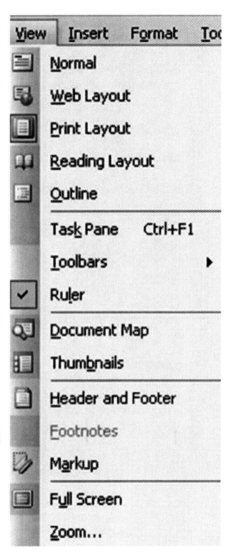

Insert – by clicking once on this you will be able to insert page numbers and pictures into your document as well as many other items.

Break allows you to start typing on a new page within the same document you are working on.

Page Numbers allows you to add page numbers to all of the pages in your document.

Date and Time allows you to insert the date and time to your document.

AutoText allows you to choose from several pieces of stored text which you can insert into your document. These range from opening lines of letters to closing lines i.e. yours faithfully etc

Picture allows you to insert either a picture which you have saved in your computer or a piece of clip art into your document.

Format – by clicking once on this you are able to change the style of your document.

Font allows you to change the size, shape and colour of your letters, numbers etc.

Paragraph allows you to change the margins and also change the line spacing.

Bullets and Numbering allows you to make lists of information using numbers or bullets

Borders and Shading allows you to put borders round paragraphs or the entire page and also allows you to shade parts of your document, therefore making it stand out.

Columns will change your document into columns – handy if you are typing a newsletter or similar.

Background allows you to change the background colour of your document but not the words.

AutoFormat will help you set up the right format for a letter or document.

Tools - by clicking once on this you will be able to access some important tools which can be used to check your document for mistakes.

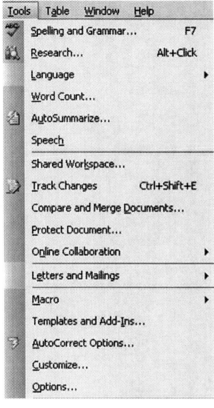

Spelling and Grammar allows you to check your document for any spelling or grammar mistakes.

Language is useful if you are typing a document in a different language as this will allow you to check for errors in that language.

Word Count allows you to count the number of words in your document - useful if you have to type a document containing a certain amount of words, you can keep a check on how many words you have done.

Letters and Mailings allows you to type envelopes, single labels or a sheet of labels and also to complete a mail merge.

Table - by clicking once on this you will be able to create a table in your document.

Draw Table allows you to create a simple table in your document.

Insert allows you to choose how many rows and columns you require in a table.

Delete allows you to remove rows and columns from your table.

Window - by clicking once on this you can open a new document window and also see all documents that you have open on the same screen.

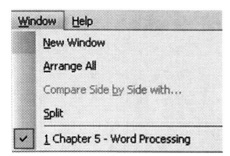

New Window allows you to open a new document window showing your current document.

Arrange All will allow you to see all the windows you have open in this program.

The last item in the list always contains the names of the documents you currently have open and are working on.

Help – by clicking once on this you should be able to find the answers to any questions you may have.

Microsoft Office Word Help will bring up a box into which you can type a question in order to get help.

Show the Office Assistant will show a virtual assistant in the shape of, among other things, a paperclip. In order to obtain help, you simply click once on the office assistant and a box will appear into which you can type your question.

Standard Toolbar

The next toolbar is the standard toolbar. Many of the items appearing on this toolbar are merely shortcuts to items found in the menus on the Menu Bar as explained above.

On this standard toolbar you have the following:

   This will open a new document.

   This will open a previously saved document.

   This will save your work.

   This will enable you to send your document via e- mail.

   This will print one copy of the document you are working on.

   This will show a preview of your document before you print it.

   This will check your document for any spelling or grammar mistakes.

Depending on which version of Word you are using, you may not have this button but it enables you to research words, finding meanings etc.

This will cut text from your document and place it on a clipboard.

This enables you to copy a piece of text and place the same text on a clipboard.

This will place the cut or copied text into the required position in your document.

This is the format brush and allows you to copy formatted text and apply it to other text.

This is your undo button and is probably the most important button on the toolbar as it enables you to go back one step so if you accidentally lose everything by pressing the wrong button, one click on this button will bring it all back.

This is the redo button which will redo the action taken by the undo button.

There are some more buttons on the end of this toolbar but these relate to inserting hypertext, tables and spreadsheets into your document which is too advanced at this stage.

Formatting Toolbar

The next toolbar down is the formatting toolbar.

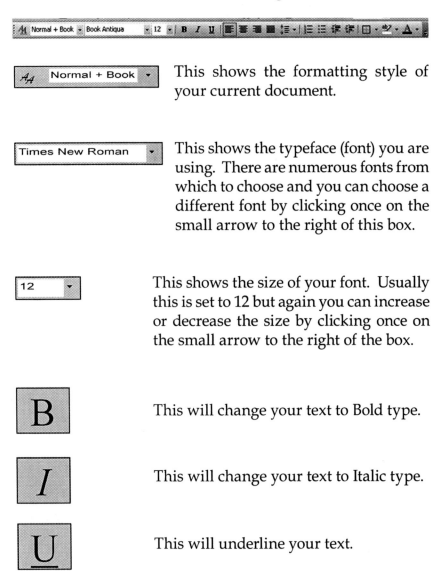

This shows the formatting style of your current document.

This shows the typeface (font) you are using. There are numerous fonts from which to choose and you can choose a different font by clicking once on the small arrow to the right of this box.

This shows the size of your font. Usually this is set to 12 but again you can increase or decrease the size by clicking once on the small arrow to the right of the box.

This will change your text to Bold type.

This will change your text to Italic type.

This will underline your text.

These are your alignment buttons. The first one will align your text to the left hand margin

The second one will centre your text.

The third one will align your text to the right hand margin.

The fourth one will justify your text which means that your text will be aligned straight against both your left and right hand margins.

This button will set the spaces between your lines. i.e. Single line spacing/double line spacing. *You may not have this button.*

If you click on this button, you can easily type a numbered list.

Again this button will enable you to type a list – this time using bullet points instead of numbers.

These are your indent buttons - the first one will decrease your indent and the second one will increase it.

This will put a border around a specific piece of text.

This button allows you to highlight text and it works in the same way as an actual highlighter pen. The small arrow to the right of the box enables you to choose a highlight colour.

 This is your font colour. Again if you click on the small arrow to the right, you can pick one of many different colours.

Getting started and typing text

When you first open Word, you will see a small line flashing on your screen in the top left hand corner, this is called the insertion point and this is where the first letter you type will appear. If you wish to start typing further down the page then just press Enter a few times and the insertion point will move down the page. As you type, the insertion point moves to the right of your text. Basically all you do is just type the required text on your keyboard and the words will appear on your screen. There is no need to press Enter/Return when you get to the end of a line, as it will automatically take you to the next line. The only time you need use the Enter/Return key is to start a new paragraph. If you wish to indent your paragraphs you can do this easily by simply pressing the tab key on your keyboard once and this will move the insertion point in about half an inch. Normally the font will be set to Times New Roman and the font size will be set to 12 but these can be easily changed.

Changing Margins

The margins in Word are pre-set and generally will not need to be changed. These are set at 1 inch at the top and bottom of the page and 1.25 inches to the left and right of the page. These will usually be fine for most documents you type but should you wish to change the margins then you will first need to click on File on the menu toolbar then click on Page Setup and adjust the margins as necessary. Click OK at the bottom of the small window and the margins will have been adjusted.

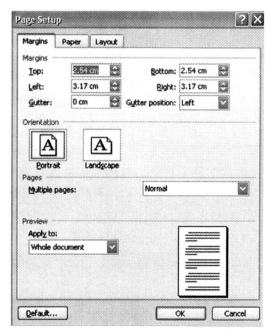

Highlighting Text (to make changes)

After you have typed your text you can edit it in many ways but before you can do this, **the text must be highlighted**. There are various ways to highlight text.

- To highlight one word double click on the word
- To highlight a sentence click once to position the insertion point at the start of the sentence and then hold down the shift key and use the right arrow key on your keyboard until the required text has been highlighted
- To highlight an entire paragraph click 3 times in the centre of the paragraph.

- Alternatively position the insertion point (by clicking with your mouse) at the start of the text you wish to change. Then click and hold down the left side of the mouse whilst dragging it over the text you wish to change.

You will now see the required text is highlighted. You can now make the changes required.

Formatting Text (changing the font, the font size, the font colour and the alignment of the text)

You must remember that in order to carry out any of the following instructions, the text must be highlighted first.

Change the font

If you wish to change the font itself, click on the small downward arrow ▼ to the right of the font box and select a new font from the drop down menu which will appear. Click once on the font required and you will see your text change to the new font.

Times New Roman ▼

Change the font size

Similarly to change the size of the font simply click on the small downward arrow ▼ to the right of the font size box and a drop down menu will appear, choose a size and click once on it – the text is now changed to the new size.

12 ▼

Change the font colour

To change the font colour, again make sure the text is highlighted, and then click on the small downward arrow▼ to the right of the font colour button which is near the end of the formatting toolbar and a palette of colours will appear. Click once on the required colour and your text will have changed colour.

A ▼

Text Alignment

You may wish to move the text around after you have typed your letter or document. For example, you may wish to centre your address at the top of the page. To do this, highlight the address and then click on the centre button on the formatting toolbar and the text will be moved to the centre of your page. Alternatively if you wish to put your address on the right of the page, you will need to click once on the right align button and the address will move to the right.

Moving Text (cut/copy/paste)

You can also move text easily in your document. For instance say you have typed a letter or document and after reading it through you realise that your letter or document would read better if a sentence you had placed at the start of a paragraph were to be moved to the end of the paragraph. This is very easy to do. Simply highlight the sentence to move, and then click once on the cut button on the standard toolbar and you will see that the text has now disappeared from your document. You should now click to position the insertion point at the place you want the sentence to be moved to and then click once on the paste button on the standard toolbar. Your text has now been moved to the new position.

You may also wish to copy a particular piece of text to a different place in your document. Again you need to highlight the required text then click on the copy button on the standard toolbar and then click to position the insertion point at the required position and click once on the paste button. Your text has now been copied and placed in the new position whilst still remaining in the original position in your document.

Deleting Text

To delete text, you will need to first highlight the required text and then simply press backspace or delete, on your keyboard. So if you just want to delete one word, double click on the word and it will be highlighted, then click once on delete or backspace. Or if you wish to delete a sentence, click to position the insertion point at the start of the sentence you wish to delete, and then hold down the shift key and use the right arrow key to highlight the text. Once highlighted, press delete or backspace on your keyboard.

If, whilst you are typing text, you notice a typing error you can quickly correct it, again by using the delete and backspace keys. If the typing error is to the right of the insertion point you should press the delete button on your keyboard but if the error is on the left of the insertion point then you should press the backspace key on your keyboard, to delete the text.

Checking your document for mistakes

Once you have completed your document, and before you decide to print it, you should first check it for spelling and grammar mistakes. You may have noticed whilst typing that some words had been underlined in red and some words may have been underlined in green. This is the computer checking your document as you go – the words underlined in red are possible spelling mistakes and the words underlined in green are possible grammar errors.

To check your document you should first click so the insertion point is at the beginning then click on the ABC spell-check button which is on the standard toolbar. If there are any errors, a small window will appear and you are given alternative words to insert, which you can accept or ignore. You will see in the following example that I have spelt underlined wrong and Word is giving me the correct spelling in the bottom pane of the small window. To easily change this mistake you just click on change and the mistake will be corrected by Word. If, however, Word picks up a word which you know is spelt right but which it is telling you is wrong, such as a place name then you simply click on ignore once and your original word will remain but the red line will disappear. Similarly with the grammar – usually words are underlined if you have put in an extra space or forgotten a comma – simply click on change or ignore to correct the error.

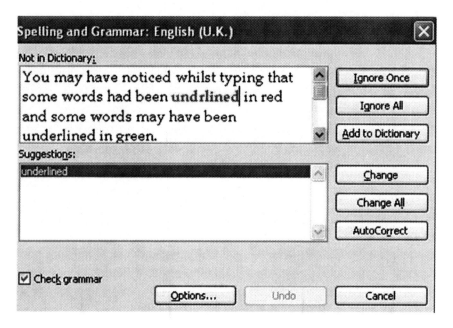

Printing

It is always a good idea to have a quick look at your document before you print it and you can do this by clicking once on the Print Preview button on the standard toolbar. You will now see how your document will look when printed. If you are not happy and wish to make changes, close the preview window and make the necessary changes.

Check the print preview once more and if you are happy you are now ready to print your document.

To do this, click once on the print button on the standard toolbar and one copy of your document will print (*make sure your printer is turned on before you click on print*).

If you wish to print more than one copy, you should click on File and then click on Print. The following window will appear. You will see that the number in the "Number of Copies" box is highlighted enabling you to overtype. Type the number of copies you require. Then click OK and your document will be printed.

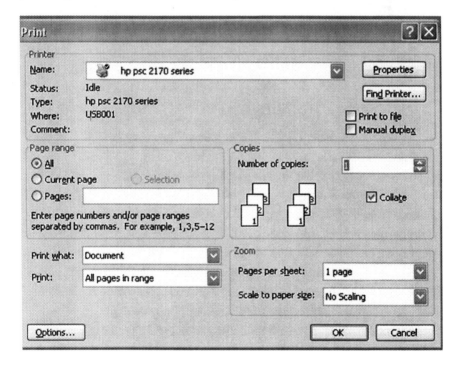

Saving your work

In order to keep a copy of your work on your computer, you must save it.

To do this click once on File and then click once on "Save As" or you can click once on the Save button on the standard toolbar. By doing either of these, the following "Save As" window will appear.

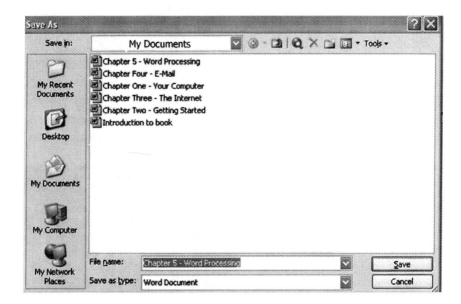

At the very top of this window, to the right of the words "Save in:" is the place where your work will be saved. This will normally say "My Documents" as this is the place where most of your work will be saved. At the bottom of the window to the right of "File name:" you will see the words have been highlighted so you can just overtype and name your document. Once you have done this, click on Save and your work will be saved. Once you have saved your work for the first time, you can save further changes by simply clicking on the Save button on the standard toolbar.

Don't forget you can keep all your saved work tidy by placing your files into folders – see Chapter Two, Section 4 – My Documents – for instructions on how to do this.

Finding your saved work

Now you have saved your work you may wish to return to it and make changes or add or delete part of it or you may just want to print an extra copy of it.

There are three ways to find any work which you have saved, as above.

1. double click on the "My Documents" icon on your desktop or
2. double click on the "Word" icon on your desktop and then click once on File on the menu bar and then click on open or
3. when you have opened Word, click on the open button on your standard toolbar.

This will bring up the following window showing all the files which you have previously saved into "My Documents".

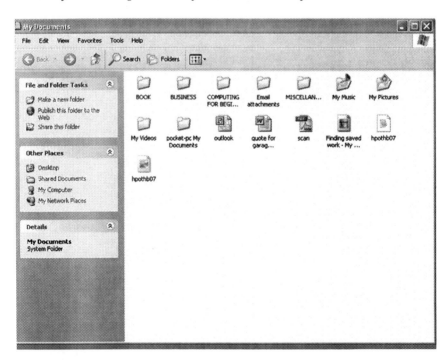

Double click on the required file and it will open so you can make any changes you require or print more copies.

Don't forget to save your work again after the changes have been made.

If you have found this book helpful, you may be interested to know that I am currently working on a second book which will take you one stage further and which will include the following:

Backing up your work to CD

<u>Word-processing</u> – the next level including inserting pictures and clipart into your work, Borders, WordArt, Columns etc

<u>Digital Photography</u> – transferring your photos to your computer, saving and printing them.

<u>Music</u> – Windows Media Player – downloading music, transferring music (burning) to CDs/MP3s.

<u>Internet</u> – how to use E-Bay.

About the Author

Lynn Manning recovered from a ten year battle against ME in 1999 and decided to re-train in computers. She spent two years studying and received various City & Guild qualifications and in 2002, set up her own computer training business offering one to one training to the beginner. This is now a very successful business and many of her customers suggested she should write a book in the same manner in which she teaches i.e. in plain English, no computer jargon, just an easy to follow manual for the beginner building on her own experience as a computer latecomer, now training newcomers and this is how she came to write this book.

Printed in the United Kingdom
by Lightning Source UK Ltd.
125436UK00001B/138/A